YOUR KNOWLEDGE HAS VALUE

Subin Raj

A historical perspective on the discovery of cellular compartments, the Organelles

The Organelles and the history behind their discovery

GRIN Verlag

Bibliografische Information der Deutschen Nationalbibliothek:

Die Deutsche Bibliothek verzeichnet diese Publikation in der Deutschen National-
bibliografie; detaillierte bibliografische Daten sind im Internet über http://dnb.d-
nb.de/ abrufbar.

Imprint:

Copyright © 2013 GRIN Verlag GmbH
Druck und Bindung: Books on Demand GmbH, Norderstedt Germany
ISBN: 978-3-656-74828-1

This book at GRIN:

http://www.grin.com/en/e-book/280878/a-historical-perspective-on-the-discovery-
of-cellular-compartments-the

GRIN - Your knowledge has value

Der GRIN Verlag publiziert seit 1998 wissenschaftliche Arbeiten von Studenten, Hochschullehrern und anderen Akademikern als eBook und gedrucktes Buch. Die Verlagswebsite www.grin.com ist die ideale Plattform zur Veröffentlichung von Hausarbeiten, Abschlussarbeiten, wissenschaftlichen Aufsätzen, Dissertationen und Fachbüchern.

Visit us on the internet:

http://www.grin.com/

http://www.facebook.com/grincom

http://www.twitter.com/grin_com

Content

Introduction

The cell was first discovered by Robert Hooke in the year 1665. The information inferred from his analysis was limited to the level of resolution offered by technology of his time (WM. TURNER, 1889). It took researchers nearly two more centuries to realize that the cellular building block of organisms itself is sub-divided into certain conserved structures. The improved technology bringing about the rise of better Microscopes and along with the development of novel staining methods were able to provide better contrast to identify and resolve sub cellular structures within the sample.

The boundary of the cell called the plasma membrane was suggested to be of lipid origin since other lipid molecules and hydrophobic substances were able to pass through it (Overton, 1889). Later in 1925, Gortnel and Grendel while working with RBCs inferred that the membrane is a bilayer of lipids based on the observations made on comparison between the surface area of the cells and volume of lipids isolated in a Langmuir trough from these cells (Adams, 2010).

The German zoologist Karl August Möbius is credited for the first usage of the term "*organula*" (Möbius, 1884; Bütschli, 1888). From this, the term organelle evolved with a set of diverse meanings, but eventually, the now widely used definition formed after the discovery of various cellular structures with surrounding membranes. Most of the Cell Biologists are of the view that the term organelle is synonymous with "cellular compartment" while others choose to limit the term organelle to include only those that are DNA-containing, having originated from formerly autonomous microscopic organisms (Keeling & Archibald, 2008).

Nucleus: One ring to rule them all

Anton von Leeuwenhoek's (1632-1723) drawing can be considered as the oldest version of the description of the nucleus. He had observed a lumen, the nucleus in the RBCs of salmon. Unlike the mammals, the rest of the vertebrae possess nuclei in their RBCs (Gerlach, 2009). Later it was again described by the botanical artist Franz Bauer, the nucleus was then studied in detail and named independently by the Scottish botanist Robert Brown in 1833

Mitochondrion: The ring of power

The earliest records on intracellular structures that probably represent mitochondria go back to the early 1840s, only a few years following the discovery of the cell nucleus. However, Altmann in 1890 was the first to recognize the ubiquitous occurrence of these structures. He had called them "bioblasts" and assumed that they were "elementary organisms" living inside cells helping in metabolic activities of the cell. Altmann would have been greatly satisfied by knowing that his idea of the symbiotic origin of mitochondria would be revived, verified and approved by the majority of the scientific community several decades later, based on similarities between mitochondria and bacteria. The name mitochondrion was introduced in 1898 by Benda, and originates from the Greek "mitos" meaning thread and "chondros" referring to the granule, based on the appearance of these structures during spermatogenesis.

In 1900, Michaelis found that the redox dye Janus Green B serves as a specific supravital stain of mitochondria. As pointed out by Palade in 1964, this stain became the feature called "official portrait" of mitochondria until 1952, when the first high-resolution electron micrographs of mitochondria were published. It is remarkable that Michaelis's active interest in the redox processes in the cell did not lead him to identify the role of Mitochondria in the cell. In fact, it took 50 years to recognize this role, until Lazarow and Cooperstein demonstrated that the specific staining of mitochondria by Janus Green B is because of their capacity to reoxidize the reduced dye by way of cytochrome oxidase. Krebs discovered the citric acid cycle in 1937 and later in 1949, Kennedy and Lehninger discovered their role in the production of ATP through oxidative phosphorylation and thus continued the story of mitochondria (Ernster & Schatz, 1981).

Chloroplast: The ring for light

Some historians have taken a discourse that the English physician Nehemiah Grew (1641–1712) as the starting point of our knowledge of chloroplast structure. His communication, "*A Discourse on the Colours of Plants*" to the Royal Society of London, UK, on May 3, 1677, elaborated how he extracted the green pigment of leaves with olive oil and noted its different colors when held up to a candle. Microbiologist Anton van Leeuwenhoek's letter number 6, dated 7 September, 1674, (translated by C. Dobell) explains how van Leeuwenhoek took water samples from the Berkelse Mere and found floating in it "some green streaks, spirally wound serpent-wise, and orderly arranged, after the manner of the copper or tin worms which distillers use to cool their liquors as they distil over. The whole circumference of these streaks was smaller than the thickness of a hair of one's head and all consisted of very small green globules joined together. It can be understood without a doubt that he had resolved the chloroplasts of *Spirogyra* and possibly, the glistening starch sheaths around its pyrenoids. Much later, Andreas Franz Wilhelm Schimper (1856– 1901), who in his later career became better known as a plant geographer, encapsulated much of the preceding work by establishing a logical terminology based on the Greek word *Plastikos*, meaning formed, or moulded. Schimper had named the green coloured ones as *Chloroplastiden*, the colorless types (usually filled with starch) as *Leukoplastiden*, and the non-green colored ones as *Chromoplastiden*. The category of plastid that we now know as amyloplasts was not separated as a subset of Schimper's "Leukoplastiden" until much later (Wise & Hoober, 2007), And so emerged the chloroplasts.

Golgi: The complex of rings

At the end of the nineteenth century Camillo Golgi, professor of General Pathology and Histology at the University of Pavia, was already internationally acclaimed for his discoveries on the structure of the central nervous system and for his description of the human malarial parasites and their developmental biology in the human blood. Aided by his black reaction method, Golgi described the complex arborisation of the dendrites, discovered the branching of the axons, analyzed several regions of the nervous system in detail and was able to publish beautiful illustrations of them. During an investigation on the structure of the spinal ganglia Golgi had noticed the occurrence of a reticular structure in the cytoplasm of the cell body of the neuron, which was clearly detached from the membrane and nucleus. Unfortunately this bizarre new structure was not constantly stainable. But Emilio Veratti (who in 1902 went on to describe the structure of sarcoplasmic reticulum working in Golgi's laboratory, was able to replicate and confirm the findings in the cell body of the fourth cranial nerve. Following this, Golgi's students succeeded in demonstrating the network in several non-neural cells like the pancreatic cell, cells from glands i.e. thyroid, pituitary epididymis, salivary glands and ovary. It became increasingly evident that this structure did not belong only to neurocytology, but was probably ubiquitous in various eukaryotic cell types. Based on the shape and intracellular location of the structure, Golgi named it "apparato reticolare interno" or internal reticular apparatus (Mazzarello, Garbarino, & Calligaro, 2009). It was Golgi and subsequently his school who defined the apparatus morphologically as an organelle, a stable component of all eukaryotic cells, and clearly differentiated it from other cellular compartments.

Endoplasmic Reticulum: The net

The endoplasmic reticulum (ER) is one of the largest, most functionally complex and architecturally varied organelle within the cell. And despite their large size and unique architecture it was one of the last organelles to be discovered. The ER was described in 1902 based on the careful observations of the Italian scientist Emilio Veratti . As a student of Golgi, Veratti used Golgi's staining procedures and found this new subcellular structure that he could verify to be distinct from the filaments of the muscle and the Golgi apparatus. Despite Veratti's careful observations and drawings he was unable to convince then contemporary scientific community that such an organelle existed. In fact, his work was disregarded and put away and thus, while research on the Golgi apparatus and other organelles was dashing forward, the research on ER was left behind. In retrospect, it is now difficult to explain how, in an era of excitement and interest in sub-cellular structures, this organelle was particularly slow to be acknowledged. In fact, it was again advance in technology that enabled the rediscovery of the ER. In 1953, Keith Porter developed techniques of electron microscopy (EM) that allowed him to resolve and observe the net-like (reticulum) structure within (endo) the cytoplasm (plastic). He therefore named the structure endoplasmic reticulum. The validation of the existence of ER came in 1954 when Porter teamed up with the father of modern cell biology, George Paladeand they were able to obtain high-resolution images and finally prove the existence of this organelle. And so, over 50 years after its initial discovery, the ER stepped into the spotlight and was accepted as a bona fide organelle, attracting curiosity thereby becoming the object of many investigations (Schuldiner & Schwappach, 2013).

And so, since its discovery more than a hundred years ago the ER has taken its place as a key protagonist in cellular organization, transport and function. Every year there arises, new fascinating facets about the functioning and regulation of ER and also the maintenance of its structure. The last decade has also demonstrated that the ER plays a central role in inter-organelle communication by hosting contact sites with all other intracellular organelles and also in transport and Post translational modification of proteins.

Lysosome: Ring of destruction

In 1949, Christian de Duve, then chairman of the Laboratory of Physiological Chemistry at the University of Louvain in Belgium, was studying the action of insulin on liver cells. He wanted to determine the localization of an enzyme called glucose-6-phosphatase inside the liver cells. He and his group knew that this enzyme played a key role in regulating blood sugar levels. They derived cellular extracts blending rat liver fragments in distilled water and centrifuging the mixture. They observed high activity of phosphatase in the extracts. However, when they tried to purify the enzyme from cellular extracts, they had an unexpected problem, they could precipitate the enzyme, but they could not redissolve the precipitate. So, instead of using cellular extracts, they decided to use a more gentle technique that fractionated the cells with differential centrifugation. This technique separates different components of cell based on their sizes and densities. The researchers ruptured the rat liver cells and then fractionated the samples in a sucrose medium using centrifugation. They succeeded in detection of the enzyme's activity in what was called as the microsomal fraction of the cell. Then serendipity entered the picture, the scientists were using an enzyme called acid phosphatase as the control for their experiments and to their surprise, the acid phosphatase activity after differential centrifugation was only 10% of the expected enzymatic activity (i.e., the activity they obtained in their previous experiments using cellular extracts). Sometime later, someone at the lab purified some cell fractions and then left them in the fridge. Five days later, after returning to measure the enzymatic activity of the fractions, they observed the phosphatase enzymatic activity level returned to the level that was expected. To ensure there was no mistake, they replicated the experiment a number of times. Each time it was seen that the results remained the same i.e. if they measured the enzymatic activity using fresh samples, then the activity was only 10% of the enzymatic activity obtained when they let the samples incubatedin the fridge for five days. Thus it was hypothesized that a membrane-like barrier limited the accessibility of the enzyme to its substrate. And incubating the samples for a few days gave the enzymes time to diffuse and act properly on the substrate when added. They described the membrane-like barrier as a "saclike structure surrounded by a membrane and containing acid phosphatase." By 1955, additional hydrolases were also discovered in these saclike structures, suggesting that these were a new type of organelle. De Duve named these new organelles as "lysosomes" reflecting their lytic nature. In that same year, Alex Novikoff from the University of Vermont visited de Duve's laboratory. An experienced microscopist, Novikoff obtained the first electron micrographs of the new organelle from samples of partially purified lysosomes. Using a stain for the enzyme acid phosphatase, de Duve and Novikoff confirmed its location in the lysosome using light and electron microscopic studies (Castro-obregon, 2014).

Peroxisome: The Ring of Cleansing

Peroxisomes which were originally called microbodies, got discovered in 1954 with the help of EM in mouse kidney cells. Since then the "Cinderella" among the sub cellular organelles, which was once considered to be a "fossil organelle" and had been regarded as the cell's "garbage pail", has experienced a remarkable rise in research studies and turned into a dynamic and metabolically active cellular compartment essential for human health and development. De Duve and Baudhuin were first to isolate peroxisomes from rat liver. Their biochemical studies led to the discovery of the co-localization of several H_2O_2-producing oxidases as well as the enzyme catalase, an H_2O_2 degrading enzyme located in the matrix of the peroxisomes. On the basis of these biochemical findings, De Duve proposed the functional term "peroxisome", which replaced the former morphological designation of "microbody" coined earlier by Rhodin (Schrader & Fahimi, 2008).

References:

Adams, M. (2010). Discovering the Lipid Bilayer Discovery of the Lipid Bilayer. *Nature Education*, *3*((9):20), 1–3.

Castro-obregon, B. S., Biotecnologia, P. D. I. De, & De, U. N. A. (2014). The Discovery of Lysosomes and Autophagy Aa The Discovery of Lysosomes The Function of Lysosomes Autophagy and Lysosomes Autophagy : A Process of Self-Digestion Autophagy in Real Time, (Straus 1954), 1–4. doi:10.1038/nrm

Ernster, L., & Schatz, G. (1981). Mitochondria: a historical review. *The Journal of Cell Biology*, *91*(3 Pt 2), 227s–255s. Retrieved from http://www.pubmedcentral.nih.gov/articlerender.fcgi?artid=2112799&tool=pmcentrez&rendertype=abstract

Keeling, P. J., & Archibald, J. M. (2008). Organelle Evolution: What's in a name? *Current Biology : CB*, *18*(8), R346–7. doi:10.1016/j.cub.2008.02.059

Mazzarello, P., Garbarino, C., & Calligaro, A. (2009). How Camillo Golgi became "the Golgi". *FEBS Letters*, *583*(23), 3732–7. doi:10.1016/j.febslet.2009.10.018

Möbius, K. (1884). The death of unicellular and multicellular animals. *Biological Central Journal*, *IV No. 13.*(No. 13.14), 389–392.

Schrader, M., & Fahimi, H. D. (2008). The peroxisome: still a mysterious organelle. *Histochemistry and Cell Biology*, *129*(4), 421–40. doi:10.1007/s00418-008-0396-9

Schuldiner, M., & Schwappach, B. (2013). From rags to riches - the history of the endoplasmic reticulum. *Biochimica et Biophysica Acta*, *1833*(11), 2389–91. doi:10.1016/j.bbamcr.2013.03.005

Wise, R. R., & Hoober, J. K. (Eds.). (2007). *The Structure and Function of Plastids* (Vol. 23). Dordrecht: Springer Netherlands. doi:10.1007/978-1-4020-4061-0

WM. TURNER. (1889). The cell theory, past and present. *J. Anat. Physiol.*, (24), 253–287.